W9-AWK-463

MATH
MADE EASY

LYNETTE FLOWERS
Contributing Writer

NIGEL J. HOPKINS
Consultant

PUBLICATIONS INTERNATIONAL, LTD.

CONTENTS

ARITHMETIC YOU'LL NEED

APPLICATIONS OF ARITHMETIC

Lynette Flowers has worked as a mathematics writer and editor for more than 35 years. She has created and developed math materials, textbooks, worksheets, and activity books for elementary, high school, and college levels.

Nigel J. Hopkins earned his Ph.D. in nuclear physics at McGill University, Montreal. He is particularly interested in applying mathematics to practical problems and is the coauthor of *The Numbers You Need* and *Go Figure!*

Illustrations: Joe Veno

Copyright © 1993 Publications International, Ltd. All rights reserved. This book may not be reproduced or quoted in whole or in part by mimeograph or any other printed or electronic means, or for presentation on radio, television, videotape, or film without written permission from:

Louis Weber, C.E.O.
Publications
  International, Ltd.
7373 North Cicero Avenue
Lincolnwood, Illinois
  60646

Permission is never granted for commercial purposes.

Manufactured in USA.
8 7 6 5 4 3 2 1
ISBN: 1-56173-738-0

INTRODUCTION

If you weren't wild about math in school and never really learned the basics . . . or if you did learn the basics once, but now they're long forgotten . . . you may find it hard to handle the calculations that come up in everyday situations. What you need is a review and refresher.

Part 1, "Arithmetic You'll Need," starts right at the beginning. It explains why numbers are written the way they are. It explains what fractions and decimals are and how they are used. You'll review how to do the four basic operations in arithmetic: adding, subtracting, multiplying, and dividing. Examples and practice exercises will make the principles clear. "Quick Tips" will show you ways you can make fast work of a calculation.

Nowadays most of us rely on our pocket calculators to take the drudgery out of arithmetic. We've included a section that explains how to use the calculator, including the percent key and the memory keys. Once you know these keys, using the calculator is even easier.

The second part, "Applications of Arithmetic," shows you how to approach the math problems you find in everyday life. Buying and selling, kitchen calculations, diets, exercising, sports, graphs and charts, operating your car, renting versus owning, conserving energy, planning home improvements, balancing your checkbook, and converting between systems of units are all covered.

Using math often concerns dollars and cents. If you work through our clear explanation of discount prices, you'll be able to calculate what you can save if you see stores offering different discounts off different prices on the same item. Here's a case in which the time you spend learning might have a real payoff.

This is a do-it-yourself book that you can work through at your own pace. You can also use it as a reference book—pick it up when you need to solve a particular problem. You could surprise yourself—with the help of this book, you might start actually enjoying doing math problems!

ARITHMETIC YOU'LL NEED

WHOLE NUMBERS, DECIMALS, AND FRACTIONS

Learning what numbers mean will improve your ability to use math. Once you understand the parts of a number, you can learn how to approximate numbers and how to calculate.

Whole numbers

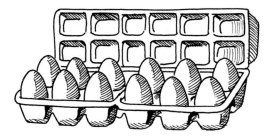

Whole numbers are used to represent quantities.

There are 365 days in a year.
The population of Chicago is 2,783,726 people.
There are 12 eggs in a dozen.
Experts recommend exercising 3 or 4 times a week.

These numbers are called whole numbers because they are not divided into smaller parts. Parts of a number are expressed in either *decimals* or *fractions*.

Decimals

The commonest use of decimals is dollars and cents; one cent is one hundredth of a dollar. In the same way, decimals represent hundredths of a whole number.

The sidewalk is 2.5 feet wide.
Earth is a minimum of 91.4 million miles from the sun.
This package of ground beef weighs 2.23 pounds.

Each digit in a number has its own value depending on the position in the number. Thus 401 is a different number from 104 or 140, even though the same digits are used. The position of the digits in our numbering system tells us how many ones, tens, hundreds, and so on are in the number.

Examples

The number 247 represents:

2 hundreds	=	200
4 tens	=	40
7 ones	=	7
The sum of these figures	=	247

The digits to the right of the decimal point represent tenths, hundredths, and so forth. If a number has no digits to the left of the decimal point, a 0 is usually written to call attention to this.

The number 356.56 represents:

3 hundreds	=	300
5 tens	=	50
6 ones	=	6
5 tenths	=	0.5
6 hundredths	=	0.06
The sum of these figures	=	356.56

Practice

1. Write the following numbers in a display like the ones above.
   a. 480 b. 7.62 c. 0.322 d. 620.054

2. What is the value of each 3 in the number 363.333?

Fractions

Fractions also express parts of numbers and are often used for measuring.

The flight lasted $2\frac{1}{2}$ hours.
The recipe calls for $\frac{3}{4}$ cup of sugar.
Typing paper is $8\frac{1}{2}$ inches wide.
The group ate $\frac{5}{6}$ of the pizza.

The bottom number of a fraction is called the denominator, and this number tells how many parts are in the whole. The top number is the numerator, and it tells how many parts of the whole are being considered. In the fraction $\frac{5}{6}$, for example, the whole is

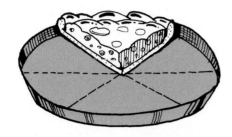

divided into 6 parts, and 5 parts of the 6 are being referred to.

Fractions can be written as decimals and vice versa. To find the decimal form of a fraction, use a calculator to divide the numerator by the denominator. For example, $\frac{3}{4}$ can be written as 0.75; $\frac{3}{8}$ is the same as 0.375. The decimal form of a fraction is usually easier to use when you are calculating.

Examples

A. Use a calculator to write $\frac{5}{8}$ as a decimal.
Press: 5 ÷ 8 =
The display is 0.625.
$\frac{5}{8}$ = 0.625

B. Use a calculator to write $\frac{1}{6}$ as a decimal.
Press: 1 ÷ 6 =
The display is 0.1666666 . . .
The 6s go on indefinitely, so you arbitrarily round off the answer at some point (see page 7).
$\frac{1}{6}$ = 0.167

Practice

3. Use a calculator to write these fractions as decimals.
a. $\frac{3}{8}$ b. $\frac{3}{2}$ c. $\frac{1}{12}$ d. $\frac{3}{5}$ e. $\frac{9}{16}$ f. $\frac{5}{3}$ g. $\frac{7}{8}$ h. $\frac{3}{20}$

Answers: 1. a. 400 + 80 **b.** 7 + 0.6 + 0.02 **c.** 0.3 + 0.02 + 0.002 **d.** 600 + 20 + 0.05 + 0.004 **2.** From the left: 3 hundreds, 3 ones, 3 tenths, 3 hundredths, 3 thousandths. **3. a.** 0.375 **b.** 1.5 **c.** 0.08333 . . . **d.** 0.6 **e.** 0.5625 **f.** 1.66666 . . . **g.** 0.875 **h.** 0.15

ROUNDING AND APPROXIMATING NUMBERS

In everyday life you use numbers in two ways—as exact numbers and as approximations. If you ask the size of the crowd at a rally, often you do not need to know the exact number; an approximation will do. Approximations also make calculating easier and can enable you to find an answer mentally.

Rounding off

QUICK TIP

The rules for rounding are:
5 and above,
round *up*
4 and below,
round *down*

Exact number	Approximation
A. 3.9 pounds	about 4 pounds
B. 411 miles	about 410 miles
C. $7.95	about $8
D. $6809.12	about $6809

Approximations are rounded numbers, or numbers in which some of the digits have been dropped. The digit to the left of the digit or digits to be dropped either remains the same or is raised by 1 according to the following rules:

1. If the last digit to be dropped is 5 or greater, the preceding digit is raised by 1.
2. If the last digit to be dropped is 4 or less, the preceding digit remains the same.

Examples

424.6 is rounded *up* to 425
424.4 is rounded *down* to 424

Depending on how accurate the approximation needs to be, you can continue this process. You can round 425 up to 430, or 424 could be rounded down to 420.

To round numbers correctly, keep in mind what value each digit in the number has. Here are the ways the number 562.439 can be rounded:

To the nearest hundredth:	562.44
To the nearest tenth:	562.4
To the nearest one:	562
To the nearest ten:	560
To the nearest hundred:	600

Practice

Use rounding to approximate these numbers.

1. The sticker price of a certain new car is $11,848.
   a. Approximate the price to the nearest hundred dollars.
   b. Approximate the price to the nearest thousand dollars.

2. The seating capacity of RFK Stadium is 55,672.
   a. Approximate the seating capacity to the nearest ten people.
   b. Approximate the seating capacity to the nearest ten thousand people.

3. The average weekly earnings of Mike, a construction worker, are $526.40.
   a. Approximate his earnings to the nearest dollar.
   b. Approximate his earnings to the nearest ten dollars.

4. The population of the United States in the 1990 census was 248,709,873 people.
   a. Approximate the population to the nearest hundred thousand people.
   b. Approximate the population to the nearest ten million people.

5. A full-grown hippopotamus can weigh 6,738 pounds.
   a. Approximate its weight to the nearest ten pounds.
   b. Approximate its weight to the nearest thousand pounds.

6. Company profits increased by 10.48% in 1992.
   a. Approximate the increase to the nearest tenth of a percent.
   b. Approximate the increase to the nearest percent.

Answers: **1. a.** $11,800 **b.** $12,000 **2. a.** 55,670 **b.** 60,000 **3. a.** $526 **b.** $530 **4. a.** 248,700,000 **b.** 250,000,000 **5. a.** 6,740 **b.** 7,000 **6. a.** 10.5% **b.** 10%

ADDING NUMBERS

Before adding a series of numbers, you line up the numbers so that digits with the same values are all in the same column. This helps you add together all digits that stand for hundreds, add together all tens digits, and so on. If the total in any column is 10 or more, then the two digits will not fit in one column. You carry the tens number to the next column to the left. For example, if the total in the ones column is 15, then you write the 5 in the ones column and carry the 1, which stands for 1 ten, to the tens column.

Carrying

QUICK TIP

You can get an approximate answer to an addition problem by rounding the numbers and adding them in your head. Your estimate will be surprisingly close.

Find the total cost of the following purchases: $7.95, $3.09, 63¢, $4, $2.65, 98¢, $1.42, and $10.80.

A. Before you add, be sure the decimal points are lined up as shown below.

$$
\begin{array}{r}
\$7.95 \\
3.09 \\
0.63 \\
4.00 \\
2.65 \\
0.98 \\
1.42 \\
+\ 10.80 \\
\end{array}
$$

B. Add the numbers column by column starting with the hundredths on the right.

$$
\begin{array}{r}
3 \\
\$7.95 \\
3.09 \\
0.63 \\
4.00 \\
2.65 \\
0.98 \\
1.42 \\
+\ 10.80 \\
\hline
2 \\
\end{array}
$$

The first sum is 32 hundredths. Write the 2 under the hundredths column and carry the 3 tenths, adding them in with the next column.

C. Add the tenths column, including the 3 tenths carried from the hundredths column.

4 3
$7.95
3.09
0.63
4.00
2.65
0.98
1.42
+ 10.80
.52

The sum is 45 tenths. Write the 5 tenths under the tenths column and carry the 4 ones to the next column.

QUICK TIP

For most lengthy addition tasks, it's faster and more accurate to use a calculator.

D. Continue adding and carrying until the computation is done.

24 3
$7.95
3.09
0.63
4.00
2.65
0.98
1.42
+ 10.80
$31.52

The total is $31.52.

Practice

Find each sum.

1. Miles driven on vacation:
   320, 285, 65, 120, 92, 247, 365

2. Bowling scores for 6 games:
   189, 174, 166, 208, 191, 188

3. Calories in food consumed during one day:
   250, 47, 120, 475, 36, 13, 150, 625, 275, 72, 183.

4. Hours worked one week:
   7.5, 8.25, 8, 9.25, 8.75, 4.5

5. Car-repair bill:
   Parts, $79.95; labor, $184.50; oil, $6.50; tax, $5.32

Answers: 1. 1,494 **2.** 1,116 **3.** 2,246 **4.** 46.25 **5.** $276.27

SUBTRACTING NUMBERS

To subtract one number from another, line up the numbers in columns and start at the right-hand column, just as you do for addition. In a column, the number you are subtracting might be greater than the number you are subtracting from. This problem is solved by *borrowing* from the next column to make the top number large enough.

Examples

QUICK TIP

It's easy to do subtraction in your head if you break the problem down into easier, more manageable calculations. To subtract $14.95 from $20, round the $14.95 up to $15. This is $5 less than $20. Then add the 5¢ to get $5.05.

How much change should you get from $20 for a compact disc costing $14.95? Subtract to find the answer.

A. Write $20 as 20.00. Line up the decimal points as you write 14.95 under 20.00.

$$\begin{array}{r} 20.00 \\ -\ 14.95 \\ \hline \end{array}$$

B. To subtract 5 from 0, 9 from 0, and 4 from 0, you need to borrow. Mentally make change for the $20 bill and think of it as 19 dollars, 9 dimes, and 10 pennies.

$$\begin{array}{r} {\scriptstyle 19\ 9\ 10} \\ 20.00 \\ -\ 14.95 \\ \hline \end{array}$$
10 minus 5 is 5.
9 minus 9 is 0.
19 minus 14 is 5.

C. Now you can subtract. Work from right to left, as in addition.

$$\begin{array}{r} {\scriptstyle 19\ 9\ 10} \\ 20.00 \\ -\ 14.95 \\ \hline 5.05 \end{array}$$

You should get $5.05 in change.

Last month Larry weighed 203 pounds. He has been on a strict diet, and when he weighed himself this morning, he weighed 186 pounds. How much weight has Larry lost?

SUBTRACTING NUMBERS

A. Subtract Larry's new weight from his old weight.

$$\begin{array}{r} 203 \\ -\ 186 \end{array}$$

B. Since you can't subtract 6 from 3 or 8 from 0, you need to borrow by "making change" for 203 just as you did for the $20 bill. Rewrite 203 as 1 hundred, 9 tens, and 13 ones.

1 9 13
$$\begin{array}{r} 203 \\ -\ 186 \\ \hline 17 \end{array}$$

13 minus 6 is 7.
9 minus 8 is 1.
1 minus 1 is 0.

Larry has lost 17 pounds.

Practice

QUICK TIP

For lengthy subtraction problems it is faster and more accurate to use a calculator.

1. How many more people does the Astrodome seat than does Wrigley Field?
   Astrodome, 54,816; Wrigley Field, 38,710

2. How much do you save by buying this refrigerator on sale?
   Regular price, $839.99; sale price, $672.50

3. How much change should you get from $40 when you buy a pair of shoes costing $32.79, including tax?

4. How much is this worker's take-home pay?
   Gross weekly salary, $540; deductions, $183.50

5. How much do you save when you buy the generic cereal?
   18-ounce box brand-name corn flakes, $2.05; 18-ounce box generic corn flakes, 89¢

6. How many more home runs did Hank Aaron hit in his career than Babe Ruth hit?
   Hank Aaron, 755; Babe Ruth, 714

7. How much do you save by buying this 26-inch stereo TV on sale?
   Regular price, $609; sale price, $485.50

Answers: 1. 16,106 **2.** $167.49 **3.** $7.21 **4.** $356.50 **5.** $1.16 **6.** 41 **7.** $123.50

MULTIPLYING NUMBERS

Multiplication is actually an abbreviated form of addition:

$$2 \times 2 = 2 + 2$$
$$3 \times 2 = 2 + 2 + 2$$
$$4 \times 2 = 2 + 2 + 2 + 2$$

and so on.

Writing out a complete set of additions for large numbers would not be practical—$378 \times 2 = 2 + 2 + 2$, and so forth, would be a very long sequence of additions. Instead we use the shorthand notation 378×2 to represent this long sequence.

To carry out multiplications in your head, you must memorize the multiplication table at least up to the $9 \times 9 = 81$ level. These relationships can be used to carry out multiplications of numbers with any number of digits.

QUICK TIP

To multiply by multiples of 10, simply add as many zeros to the product as there are in the multiple of 10 you are using. For example, to multiply by 10, add one zero to the product. To multiply by 100, add two zeros, and so on.

Multiplication Table

1	2	3	4	5	6	7	8	9
2	4	6	8	10	12	14	16	18
3	6	9	12	15	18	21	24	27
4	8	12	16	20	24	28	32	36
5	10	15	20	25	30	35	40	45
6	12	18	24	30	36	42	48	54
7	14	21	28	35	42	49	56	63
8	16	24	32	40	48	56	64	72
9	18	27	36	45	54	63	72	81

Example

Multiply 98 by 7.
The steps to take are as follows:

A. Beginning on the right, multiply 7 by 8 and write the answer 56 as shown with the 6 in the ones column and the 5 as a smaller figure "carried" into the tens column and written above the 9.

$$
\begin{array}{r}
5 \\
98 \\
\times\, 7 \\
\hline
6
\end{array}
$$

B. Next multiply 7 by the 9 in the tens column. Before writing down the answer 63, add the carried 5 to it and write the answer 68 in the places shown:

$$\begin{array}{r} 98 \\ \times\ 7 \\ \hline 686 \end{array}$$

Long multiplication

The same basic procedure is used for more complicated cases.

Multiply 98×37.

A. The first steps are the same as before; multiply 98 by the 7 in the ones column and write down the answer.

$$\begin{array}{r} 98 \\ \times\ 37 \\ \hline 686 \end{array}$$

B. Next multiply 98 by the 3 in the tens column. Write the answer below the previous result but moved one place to the left, so that the digits line up with the tens column.

$$\begin{array}{r} 2 \\ 98 \\ \times\ 37 \\ \hline 686 \\ 294 \end{array}$$

C. The final step is to add the two partial products:

$$\begin{array}{r} 98 \\ \times\ 37 \\ \hline 686 \\ 294 \\ \hline 3626 \end{array}$$

The same procedure can be extended for multiplying numbers of any size, but it is tedious work and can be done more efficiently and more accurately on a calculator.

QUICK TIP

In converting a larger unit of measure to a smaller one, it helps to remember there will be more of the smaller units. Since a kilometer is shorter than a mile, a conversion of miles to kilometers will result in a larger number.

Multiplying decimals

To multiply numbers with decimals, ignore the decimal points until you have finished multiplying. Then place the decimal point in your answer according to this rule: The number of digits to the right of the decimal point in the answer is the same as the sum of digits to the right of the decimal point in the two numbers you are multiplying. For example:

$0.89 \times 0.21 = 0.1869$.
Count the number of digits to the right of the decimal point in the two numbers you are multiplying. In this case the number is four. The answer will have four digits to the right of the decimal point.

If the original numbers are accurate only to two places of decimals, the answer also will be accurate only to the same two places. Any additional digits are not "significant." The answer should always be rounded to include only significant digits.

Example

$1.23 \times 4.56 \times 7.89 = 44.253432$

The calculator shows six places of decimals as it should, but the final four are not significant and the correct answer should be 44.25, rounded to two places of decimals like the numbers being multiplied.

Uses of multiplication

One of the most common uses of multiplication is converting units of measure from one system to another.

Marathon races are 26.22 miles long, and one mile equals about 1.6 kilometers. About how many kilometers long is a marathon?

Multiply 26.22 by 1.6 to find the answer. Use a calculator. Press: $26.22 \times 1.6 =$
The answer is 41.952, or about 42 kilometers.

Practice

QUICK TIP

To multiply a decimal by multiples of ten, simply move the decimal point to the right, one digit for each zero in the multiplier.

1. A car averages 32.4 miles per gallon of gas. How far can the car travel on 16 gallons of gas?

2. Roast beef is priced at $3.70 per pound. How much does a 5.45-pound roast cost?

3. One tablespoon of butter has 12 fat grams. A stick of butter equals 8 tablespoons, and a pound of butter equals 4 sticks. How many fat grams are in one pound of butter?

4. On a city map, one inch represents 0.75 miles. A major street is 22.5 inches long on the map. How many miles long is the street?

5. Weekly union dues are $4.75. How much are union dues for one year?

6. Curtain material costs $6.95 per yard. How much would 20 yards cost?

7. One inch is the same length as 2.54 centimeters. How many centimeters long is a 1-foot ruler?

8. Banquet tickets are priced at $13.75. How much will 100 tickets cost?

Answers: 1. 518.4 miles **2.** $20.165, rounded to $20.17 **3.** 384 **4.** 16.875 **5.** $247 **6.** $139 **7.** 30.48 centimeters **8.** $1,375

DIVIDING NUMBERS

When you divide two numbers, you are finding out how many times one number is contained in the other. The number you divide by is called the divisor, the number being divided is called the dividend, and the answer is called the quotient. The process of division is taking out groups of the divisor from the dividend in several steps.

Example

There are 64 players at a bridge tournament. How many tables of 4 players are there?

A. The problem is to find the number of groups of 4 in 64. Write the divisor and the dividend separated by a division bracket.

$$4\overline{)64}$$

B. As a first step, estimate the number of tables. Since 10 tables of 4 would be 40 players and 20 tables of 4 would be 80 players, the number of tables is somewhere between 10 and 20. Use 10 as a partial quotient. Subtract 10 tables of 4.

QUICK TIP

Division is the inverse of multiplication.

Division: 24÷6=4

Multiplication:

           24=6×4

$$\begin{array}{r} 10 \\ 4\overline{)64} \\ \underline{40} \\ 24 \end{array}$$

Write 10 above the 64, lining up the tens digit and ones digit with the digits in 64. Multiply 10 times 4 and write the result, 40, under the 64, lining up the digits once again.

C. Subtract 40 from 64. After taking out 10 tables of 4, 24 people are left. Next divide 24 by 4. Since 6×4=24, 24÷4=6. These 6 additional tables need to be added to the partial quotient of 10.

After subtracting out 6 tables of 4, the remainder is 0. The division is completed.

There are 10+6=16 tables of bridge players.

When the divisor has only one digit, you can use this "short division" method. If you are dividing by two or more digits, you need to use "long division," which involves many shortcuts and combines many steps. Here is how the example just completed would be worked using long division.

$$\frac{16}{4\overline{)64}}$$
$$\underline{4}$$
$$24$$
$$\underline{24}$$
$$0$$

6÷4=1, with a remainder of 2. Write the 1 over the 6 and multiply 1×4=4, writing the 4 under the 6. Subtract 4 from 6 and bring down the next digit in the dividend to get 24. Divide and subtract again. 24÷4=6. Write the 6 over the 4; multiply 6×4=24 and write the 24 under the first 24. The remainder is 0.

Dividing decimals

> ### QUICK TIP
> To get a fast estimate in your head, change the numbers to something evenly divisible. If you want to know the answer to $4,500 divided by 12, recall that 4800÷12=400. The answer will be a little under $400. In some situations that will be close enough.

2.74 pounds of boneless chicken breasts cost $4.11. What is the price per pound?

To find the price per pound, divide 4.11 by 2.74.

A. To eliminate the decimal point in the divisor, multiply both numbers by 100. Multiplying both numbers by the same number will not change the answer; 411 divided by 274 is exactly the same as 4.11 divided by 2.74. Next write 411 as 411.00 so that the answer will be a decimal showing the price in dollars and cents.

$$274\overline{)411.00}$$

B. Estimate the answer. Since 2×250=500 and 274 is more than 250, you can tell 274 divides 411 only once.

$$\frac{1}{274\overline{)411.00}}$$
$$\underline{274}$$
$$137$$

Write 1 in the quotient over the ones digit in 411. Multiply 1×274 and write 274 under 411. Subtract to get a remainder of 137.

C. Bring down the next digit, 0, to produce 1370. Divide again by 274, estimating as you did before. Since 4×300=1200, the next digit is likely to be 5.

$$\frac{1.5}{274\overline{)411.00}}$$
$$\underline{274}$$
$$1370$$
$$\underline{1370}$$
$$0$$

Bring the decimal point up into the quotient and write the 5 over the first 0. Multiply 5×274= 1370, with a remainder of 0.

D. Bring down the last 0 and again divide by 274.

$$\begin{array}{r} 1.50 \\ 274\overline{)411.00} \\ 274 \\ \hline 1370 \\ 1370 \\ \hline 00 \end{array}$$

274 divides 00 zero times with a remainder of 0. Write 0 in the quotient over the last 0. (Not all division problems come out even; in some cases a remainder smaller than the divisor will be left over.)

Since 411.00 (the dividend) has two digits to the right of the decimal point, the quotient will also have two digits to the right of the decimal point. $411.00 \div 274 = 1.50$, which is the same as $4.11 \div 2.74 = 1.50$.

The chicken breasts are on sale for $1.50 per pound.

Practice

1. The annual leasing fee for a car is $4,500. What is the monthly leasing rate?

2. In 1991 Michael Jordan scored 2,580 points in 82 games. What is the average number of points he scored per game?

3. Doughnuts cost $4.20 per dozen. How much does one doughnut cost?

4. One gallon is equivalent to 128 fluid ounces. How many 8-ounce cups are in one gallon?

5. Gasoline is priced at $1.30 per gallon. How many gallons of gas will you get for $20?

6. You bicycle 6 days straight for a total of 150 kilometers. What is your daily average?

7. Last week you worked 38 hours. Your gross pay was $585.20. What is your hourly rate of pay?

Answers: 1. $375 **2.** 31.46 **3.** 35¢ **4.** 16 **5.** 15.38 **6.** 25 **7.** $15.40

FINDING A FRACTION OF A NUMBER

To find a fraction of a number, you multiply the number by the numerator (top number) of the fraction and divide by the denominator (bottom number).

Examples

QUICK TIP

When the fraction's top number is 1, you can find the answer in one step simply by dividing the number by the fraction's denominator.

If you are making $\frac{3}{4}$ of a recipe that calls for 2 cups of milk, how much milk should you use? Find $\frac{3}{4}$ of 2.

A. First multiply 2 by the numerator.
$2 \times 3 = 6$

B. Then divide by the denominator.
$6 \div 4 = 1\frac{2}{4} = 1\frac{1}{2}$

You should use $1\frac{1}{2}$ cups of milk.

To find a fraction of a fraction, you multiply the two numerators and multiply the two denominators.

If you are making $\frac{1}{2}$ of a cookie recipe that calls for $2\frac{1}{2}$ cups of flour, how much flour do you need?

A. Find $\frac{1}{2}$ of $2\frac{1}{2}$. First convert $2\frac{1}{2}$ to a fraction. Since 2 cups is the same as four $\frac{1}{2}$ cups, $2\frac{1}{2}$ cups is the same as $\frac{5}{2}$ cups. So you want to find $\frac{1}{2}$ of $\frac{5}{2}$.

B. $\frac{1}{2}$ of $\frac{5}{2} = \frac{1}{2} \times \frac{5}{2} = \frac{1 \times 5}{2 \times 2} = \frac{5}{4}$
Since $\frac{5}{4} = 1\frac{1}{4}$, you need $1\frac{1}{4}$ cups of flour.

Practice

1. A gas tank that holds 20 gallons is $\frac{3}{4}$ full. How much gas is in the tank?

2. One-fifth of Betty's monthly income of $3,000 goes for rent. How much rent does she pay?

3. One kilometer is about $\frac{6}{10}$ of a mile. Which is faster, a speed of 65 miles per hour or a speed of 115 kilometers per hour?

Answers: **1.** 15 gallons **2.** $600 **3.** 115 kilometers per hour

FINDING PERCENTAGES

"Percent" means "per hundred." If you read that 12 percent of the voters are undecided, that means out of every 100 voters, 12 are undecided. If the state sales tax is 6 percent, that means for every dollar, or 100 cents, the tax is 6 cents.

Converting fractions to percentages

Fractions can be converted to percentages. Just divide the top number by the bottom number, multiply by 100, and attach a percent sign.

A. If $\frac{1}{4}$ of the students have art class today, what percentage have art class?

Divide $1 \div 4 = .25$. Multiply by 100 by moving the decimal point over two digits. 25% of the students have art class today.

B. If the sale price is $\frac{2}{3}$ of the regular price, what percentage of the regular price is the sale price?

Divide $2 \div 3 = .6666 \ldots$ Round off the answer to find that the sale price is 67% of the regular price.

When percentages are used in calculations, they are written as decimals, or parts per hundred.

$40\% = 0.40$
$1\% = 0.01$
$8\frac{1}{4}\% = 0.0825$
$125\% = 1.25$

Finding the amount

Most calculations that use percentages fall into two categories. Either you know the percentage and you want to find the amount, or you know the amount and want to find what percent that amount is of the whole.

When you want to find the amount, you are trying to find:

What is X% of Y

By substituting an equal sign for "is" and a question mark for "what," you begin to create an equation:

? = **X% of Y**

Substitute a multiplication sign for "of" and you have the formula.

? = **X% × Y**

To find the amount, multiply the percentage times the number.

Examples

Every item in the store is discounted 30%. What is the discount on a lawn mower regularly priced at $569.99?

A. Round 569.99 up to 570. You want to know what is 30% of 570. Convert 30% to the decimal 0.30.

? = **X% × Y**
? = **0.30 × 570**

B. Multiply 570 by 0.30.
The answer is 171. The discount on the lawn mower is $171.

If the state sales tax is 6.25%, what is the amount of tax on a car costing $11,450?

A. You want to know what is 6.25% of 11,450. 6.25%=0.0625. The formula is:

? = **X% × Y**
? = **0.0625 × 11,450**

Use a calculator. Press: .0625 × 11450
The display reads 715.625. Round the answer to the nearest hundredth to express it in dollars and cents.

The sales tax on the car is $715.63.

Finding the percentage

In the foregoing examples, you knew the percentage and you wanted to find the amount. In another com-

QUICK TIP

When trying to calculate using a price like $39.95, round the answer up to the nearest dollar. $40 is easier to compute with, and the extra 5¢ won't have a great effect on the answer.

mon type of percentage problem, you know the amount and want to find what percentage that amount is of the whole. You can use the same formula as before, except that now the question mark (representing "what") is in a different place.

$$X = ?\% \times Y$$

Now we must isolate the question mark on one side of the equal sign. This can be done by dividing both sides by Y. (Dividing both sides of an equation by the same number does not change the result.) The new formula is:

$$X \div Y = ?\%$$

In other words, you divide the amount by the whole to find the percentage.

Examples

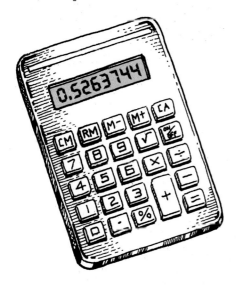

875,430 people voted in the last mayoral election. The winning candidate received 460,804 votes. What percentage of all the votes did the winning candidate receive?

A. 460,804 is what percentage of 875,430? The formula is:

$$X \div Y = ?\%$$
$$460,804 \div 875,430 = ?\%$$

B. Use a calculator. Press: 460804÷875430
The display is 0.5263744.

C. To write 0.5263744 as a percentage, multiply by 100. Move the decimal point two places to the right.
0.5263744=52.63744%

The winning candidate received about 52.6% of the votes.

Last year's rent was $480 per month. This year's rent is $520 per month. This year's rent is what percentage of last year's rent?

A. You can't assume you always divide the smaller number by the larger number. This time the question is: 520 is what percentage of 480?

520÷480 = ?%

Use a calculator. Press: 520÷480
The display is 1.0833333.

B. Round the quotient to 1.083. Write 1.083 as a percentage by multiplying by 100. Move the decimal point two places to the right.

1.083=108.3%

This year's rent is 108.3% of last year's rent.

Practice

QUICK TIP

To find 10% of any number, just move the decimal point one digit to the left. Computing a 15% tip is easy. Find 10% of the bill, divide that figure by 2 to get 5%, and add the two amounts together.

1. How much is a 15% tip on a restaurant bill of $16.50?

2. For tax purposes, a $125,000 computer system is depreciated 16.7% of its original cost each year. How much is the system depreciated each year?

3. This year's income is 125% of last year's income, which was $15,640. What is this year's income?

4. What percentage of Ken's gross pay is deducted each month?
   Gross monthly pay, $1,800; monthly deductions: $540

5. The finance charge on a $245 credit-card balance is $4.41. What percentage of the balance is the finance charge?

6. A company with 365 workers laid off 75 of these workers. What percentage of their workers did the company lay off?

Answers: **1.** $2.475, rounded to $2.50 **2.** $20,875 **3.** $19,550 **4.** 30% **5.** 1.8% **6.** 20.55%

Some problems require you to use two or more arithmetic operations to arrive at the answer. The *average* of a group of amounts, for example, is the sum of all the individual amounts divided by the number of amounts.

Examples

What is the average of these ten test scores?

64, 66, 70, 72, 78, 83, 87, 87, 91, 93

A. First add all the scores.

64+66+70+72+78+83+87+87+91+93=791

B. Divide the total of the scores by the number of scores.

791÷10=79.1

The average of the test scores is 79.1.

You may need to both multiply and add in order to find the total cost of several purchases.

What is the total cost of the following purchases if sales tax is 6%?

Two printed T-shirts costing $14.95 apiece
One pair of sweat pants costing $8.79
One sweatshirt costing $12.50

A. First multiply to find the cost of two T-shirts.

$14.95×2=29.90

B. Now add to find the total for the purchases before tax.

$29.90+8.79+12.50=$51.19

C. Now multiply the total for the purchases by 0.06 to find the tax.

$51.19×0.06=$3.07

D. Add the tax to the cost of the purchases.

$51.19+3.07=$54.26

The total with tax is $54.26.

Example

During a 6-game basketball championship series, the leading scorers on the opposing teams scored the following numbers of points.

Player A: 27, 32, 18, 26, 38, 24
Player B: 19, 31, 40, 16, 25, 25

Which player had the better scoring average for the series?

A. Find each average.

Player A: 27+32+18+26+38+24=165÷6=27.5
Player B: 19+31+40+16+25+25=156÷6=26

B. Since 27.5 is greater than 26, player A had the better average.

Practice

1. A restaurant bill for two meals is $34.40. How much change will you get from $50 if you leave a 15% tip?

2. A house bought for $75,000 in 1985 sold for $110,000 in 1992. The real estate agent's commission was 7% of the sale price. How much was the seller's profit?

3. Which costs less: a $530 air conditioner on sale at 25% off, or a $479 air conditioner with a manufacturer's rebate of $100?

4. Marcie kept track of her daily calorie intake for one week. Here are the results: 1,350, 1,375, 1,325, 1,400, 1,300, 1,340, 1,380. What was her average calorie intake per day?

Answers: **1.** $10.44 **2.** $27,300 **3.** The air conditioner with rebate **4.** 1,353 .

USING SPECIAL CALCULATOR KEYS

Calculators can do more than simple addition, subtraction, multiplication, and division. Using the percent key and the memory key, you can easily do calculations requiring several steps. These special keys work differently on various calculators. After pressing the keys shown in these examples, if your calculator does not give these results, refer to the manual for your particular model.

Percent key

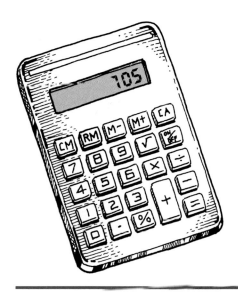

Using the percent key, you can not only find a percentage of a number but also simplify addition and subtraction of percentages. The step of placing the decimal point is done automatically. The percent key usually works in one of two ways. You can check to see which way your calculator works.

How much is a 25% discount on a regular price of $140?

To find the discount, press these keys:

140 × 25 % = OR 140 × 25 %

The display reads 35; the discount is $35.

To subtract the discount from the regular price automatically, press:

140 − 25 % = OR 140 × 25 % −

The display reads 105.

Example

After a 13% increase on a monthly rent of $750, what will the new rent be? To add the increase, press:

750 + 13 % = OR 750 × 13 % +

The display reads 847.5; the new monthly rent is $847.50.

Example

42 of the 60 people in Beth's exercise class are women. What percent of the people in the class are women?

To find the percent, press these keys.

42 ÷ 60 % = OR 42 ÷ 60 %

The display is 70. 70% of the people in the class are women.

Practice

Use the percent key.

1. A quarterback attempted 30 passes in one game. 12 of these passes were completed. What percent of his passes were completed?

2. What is the interest on a credit-card balance of $785.15 if the interest rate is 1.8% per month?

3. A broiled chicken leg with the skin on has 17 grams of fat. A broiled chicken leg without skin has 9 grams of fat. The fat in the skinned leg is about what percent of the fat in the leg with skin?

Memory keys

The memory keys are useful for doing calculations that take several steps. Many calculators have these memory keys.

CM or MC or AC: This key clears the memory of any numbers that were previously stored in it.

M+: This key adds to the memory the number in the display or the result of the current calculation.

M−: This key subtracts from the memory the number in the display or the result of the current calculation.

RM or MR or RCL: This key recalls what is in the memory and displays it.

Example

What is the total interest earned in one year on the following certificates of deposit?
$6,500 paying an annual rate of 4.8%
$15,800 paying an annual rate of 4.25%

A. First find the interest on the $6,500 CD and store this amount in the memory.

Press: CM 6500 × 4.8 % M+

The M+ key adds the display, 312, to the memory.

B. Then find the interest on the $15,000 CD and add it to the amount already in memory. This time you do *not* want to clear the memory, because the result of your first calculation is in there.

Press: 15800 × 4.25 % M+

The display is 671.5, the result of the second calculation. Pressing the M+ key adds it to what is already in the memory. To retrieve the total in the memory, press: RM

The display is 983.5. Total interest is $983.50.

> ### QUICK TIP
> **Before starting a calculation using the memory keys, always press the CM key to make sure the memory is empty.**

Practice

Use the memory keys of your calculator.

4. Find the total for the following purchases at the farmers' market.
   3 pounds of plum tomatoes at 29¢ per pound
   1.5 pounds of green beans at 39¢ per pound
   4 pounds of new potatoes at 19¢ per pound
   2 pounds of large peaches at 49¢ per pound
   2 boxes of blueberries at 69¢ per box

5. The interest rate for a loan at a credit union is 0.9% per month on the unpaid balance. How much total interest is owed on two loans, one with a balance of $11,500 and one with a balance of $7,825?

Answers: **1.** 40% **2.** $14.13 **3.** About 53% **4.** $4.58 **5.** $173.93

FINDING AREA AND VOLUME

You find the area of a rectangle by multiplying its length by its width.

Area

An office advertised for rent is 20 feet long and 12 feet wide. Express the size of the office in square feet.

20×12=240

The size of the office is 240 square feet.

You can use this basic method to find the area of an irregularly shaped region. All you need to do is break down the region into rectangles.

Example

The Pauls plan to buy new wall-to-wall carpeting for their L-shaped living/dining room. How many square yards of carpeting will they need?

A. To find the area, first separate the region into two rectangles, as shown.

B. Now find the area of each rectangle.

20 ft. × 16 ft.

8 ft. × 10 ft.

Rectangle 1: 20×16=320 square feet
Rectangle 2: 8×10=80 square feet

C. Find the sum of the two areas.

320+80=400

D. Since 9 square feet=1 square yard, divide 400 by 9 to find the number of square yards.

400÷9=44.4 square yards

The Pauls will need about $44\frac{1}{2}$ square yards of carpeting.

Volume

When you want to know how much a storage box or a freezer chest will hold, you find the volume of the container. To find the volume of a container, you multiply

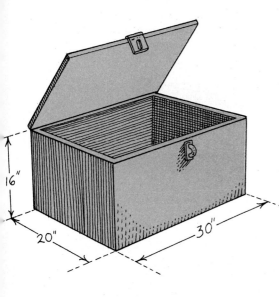

the area of the base by the height. Volume is given in cubic units.

The storage chest in Betty's pickup truck is 30 inches long by 20 inches wide by 16 inches tall. What is the volume of the storage box in cubic feet?

A. First find the area of the base.

20×30=600 square inches

B. Now multiply the area of the base by the height to get the volume in cubic inches.

600×16=9600 cubic inches

C. Since 1,728 cubic inches=1 cubic foot (see page 63), divide 9,600 by 1,728 to find the number of cubic feet.

9600÷1728=5.56

The volume of the storage box is about 5.5 cubic feet.

Practice

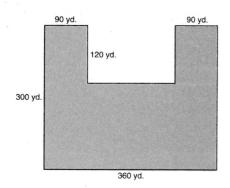

1. Find the area in square yards of this U-shaped parking lot.

2. What is the capacity in cubic feet of a storage box that measures 48 inches long, 24 inches wide, and 20 inches high?

3. The inside liner of a freezer chest measures 60 inches long by 24 inches wide by 27 inches high. What is the capacity of the freezer chest in cubic feet?

Answers: 1. 86,400 square yards **2.** 13.3 cubic feet **3.** 22.5 cubic feet

ESTIMATING AT THE SUPERMARKET

If you are at the supermarket with a limited amount of money, you might want to make a quick estimate of how much the items you already have in your cart are going to cost. One way to do this is by rounding the cost of each item so that you can add in your head.

Example

You have $25 with you. Do you have enough to pay for the following items in your cart?

Sliced roast beef	$2.73
Box of cereal	1.89
Quart of milk	0.69
Loaf of bread	1.09
Box of raisins	1.39
Cottage cheese	1.59
Frozen yogurt	2.19
Carton of cola	2.09
Head of lettuce	0.79
Mayonnaise	1.69

One way to do a quick estimate is to round all amounts up to the next dollar. This rough estimate will be higher than the actual total, thus allowing for taxes. This quick estimate gives:

3+2+1+2+2+2+3+3+1+2=21

Since $21 is a high estimate, you know that you have enough money with you.

Now suppose you want to add a container of laundry detergent costing $4.19 to your cart. Will you still have enough to pay for all the items?

You can make a closer estimate by rounding all amounts to the nearest dollar. This means that amounts less than 50¢ are ignored and amounts 50¢ or greater are rounded up to the next dollar. The new estimate including laundry detergent is:

3+2+1+1+1+2+2+2+1+2+4=21

This estimate of $21 should be fairly close to the actual amount, so you should have enough even after adding sales tax.

Practice

1. You have $50 to shop for a holiday dinner. Roughly estimate the total for these items already in your cart. Is the total under $50?

Turkey	$8.80
Cranberries	0.79
Pumpkin pie	3.09
12 cans of cola	4.65
Butter	2.19
Peas	1.59
Cream	1.89
Sweet potatoes	1.19
Stuffing mix	2.59
Walnuts	2.29
Dinner rolls	1.89
Carrots	0.89
Coffee	2.59
Sparkling wine	8.49

2. Is $50 enough if you include a $3.49 roll of film with the item listed in Exercise 1? Estimate by rounding to the nearest dollar.

Answers: 1. Yes; rough estimate is $49. **2.** Yes; estimate is $48 including the roll of film.

FINDING THE BEST BUY

Many grocery products—cereal, milk, detergent, and so on—come in packages of several sizes. To find which size gives the most for the money, you need to convert the package price to the price per unit. Finding unit prices requires dividing decimals and converting from one unit of measure to another.

Examples

A 6-pack of 12-ounce cans of grape soda is on sale for $1.39. The regular price for a 12-pack of cans is $3.19. Which is the better buy?

The unit price for the soda is the price per can. To find the price per can, divide the total price by the number of cans.

Price per can for the 6-pack: 1.39÷6=23.2¢
Price per can for the 12-pack: 3.19÷12=26.6¢

The 6-pack on sale is the better buy.

Brand-name paper towels are 89¢ for a single roll and $3.30 for a 6-roll package. Generic paper towels are 49¢ a roll. Which is the best buy?

Find the unit price for the 6-roll package of brand-name towels:

$3.30÷6=55¢

Generic towels have the lowest unit price. Are they the best buy? Not necessarily; the brand-name towels might be stronger and more absorbent. Suppose 2 sheets of the brand-name towels can do the same clean-up job as 3 sheets of generic towels.

The working cost of the 6-roll package of brand-name towels is 2×55¢=1.10.

The working cost of generic towels is 3×49¢=1.47.

In this case, the 6-roll package of brand-name towels is the best buy, the roll of generic towels is the second-best buy, and the single roll of brand-name towels comes in last.

Practice

1. Which is the better buy?
   17-ounce size of oat bran cereal for $2.59
   13-ounce size of oat bran cereal for $1.89

2. Which is the better buy?
   1½ dozen oranges for $2.79
   oranges at 15¢ apiece

3. Which is the better buy?
   36-exposure roll of film for $5.49
   24-exposure roll of film for $3.19

4. Which is the better buy?
   ½ gallon of frozen yogurt for $4.29
   3 pints of frozen yogurt for $3.00
   (See page 63 for equivalent measures)

5. Which is the better buy?
   8.2-ounce tube of toothpaste for $2.69
   6.4 ounce tube of toothpaste for $2.49, with a 30¢-off coupon

6. Which is the best buy?
   5 pounds of ground beef at Katy's Market for $4.45
   3 pounds of ground beef at Lin's Groceries for $2.46
   1.5 pounds of ground beef at Pete's Meats for $1.64

7. Which is the best buy?
   1 quart of milk for 79¢
   ½ gallon of milk for $1.58
   1 gallon of milk for $2.99
   (See page 63 for equivalent measures)

8. Which is the best buy?
   2 gallons of liquid laundry detergent for $10.95
   128 ounces of liquid laundry detergent for $8.79
   64 ounces of liquid laundry detergent for $5.14

Answers: 1. 13-ounce size **2.** oranges at 15¢ apiece **3.** 24-exposure roll **4.** 3 pints of yogurt **5.** 8.2-ounce tube **6.** 3 pounds at Lin's **7.** 1 gallon of milk **8.** 2 gallons of detergent

FINDING PERCENTAGE DISCOUNTS

Radio and television, newspapers, and catalogues constantly show ads for sales that offer discounts on merchandise. To compare prices accurately, you need to calculate the actual dollar amounts the percentage discounts represent. This requires computing with decimals and percents.

Examples

A microwave oven that regularly sells for $239 is on sale at a 30% discount. What is the amount of the discount? What is the discounted price?

A. To find the amount of discount, find 30% of $239. Remember that 30%=0.30. ? = 0.30 × 239. The discount is $71.70.

B. Subtract the amount of the discount to find the discounted price.

239.00 − 71.70=167.30

The discounted price is $167.30.

At one sports-equipment store, an exercycle that regularly sells for $399.99 is discounted 20%. At another store, the same exercycle regularly sells for $349.99 and is discounted 10%. At which store will you get the exercycle for a lower price?

A. At the first store, the discounted price is 100%−20%=80% of the regular price. Find 80% of $400.

400×0.80=320.00

B. At the second store, the discounted price is 100%−10%=90% of the regular price. Find 90% of $350.

350×0.90=315.00

You will get the exercycle for $5 less at the second store.

QUICK TIP

If you want to know the discounted price but don't care about knowing the discount, subtract the discount percent from 100% and compute with that number.

Practice

1. What is the discount percent on pizzas that are sold two for the price of one?

2. A jacket that regularly sells for $89.50 is discounted 25%. What is the amount of discount?

3. A $519 camera is discounted 30%. What is the sale price of the camera?

4. In one store a $699 refrigerator is discounted 15%. In another store the same refrigerator is selling at $100 off the $699 price. At which store is the refrigerator cheaper?

5. A VCR priced at $495 in one electronic store is on sale at 15% off. The same VCR is priced at $549.99 at another store and is on sale at 20% off. At which store will you get the lower price on the VCR?

6. An Oriental rug originally priced at $2,500 is put on sale at 20% off. One month later, still unsold, the rug is discounted an additional 15% of the already discounted price. What is the price of the rug after the second discount?

7. If the rug described in problem 6 had been discounted 35% of its original price, what would the sale price of the rug have been?

8. Explain why the sale prices for problems 6 and 7 are not the same.

9. A car with a list price of $15,500 is discounted 10% by the dealer. What will the buyer pay if the state sales tax is 6%?

10. A $3,350 computer is on sale for $2,500. Is this sale price lower or higher than a sale price that takes a 20% discount off the regular price?

Answers: **1.** 50% **2.** $22.38 **3.** $363.30 **4.** At the store with the 15% discount **5.** At the store with the 15% discount **6.** $1,700 **7.** $1,625 **8.** In problem 6, the 15% is figured on $2,000; in problem 7, the 15% is figured on $2,500. **9.** $14,787 **10.** The sale price of $2,500 is lower.

BUYING ON CREDIT

Installment buying is a purchase plan in which a customer pays only a part of the price of an item at the time of purchase. This amount is the down payment. The remainder of the purchase price is paid in equal installments at regular intervals—for example, once a month. A fee is normally added to cover the cost of extending credit.

Installment plan

A color television set can be purchased for $889.95 in full or for 4 equal payments of $250. How much more does the TV cost under the installment plan? What percent interest is this amount?

The total paid in the 4 installments is 4×$250, or $1,000. The difference between $1,000 and $889.95 is $110.05. Paying for the TV on the installment plan costs $110.05 more.

To find the percentage, divided $110.05 by the purchase price.

$110.05 ÷ $889.95 = 0.1237 = 12% interest

Credit cards

Another way to spread out the payments is to use a credit card and pay a part of the balance each month. When a credit-card user does not pay the full balance on the bill, interest called a finance charge is computed on the average daily unpaid balance. The amount of the finance charge varies among credit cards. Commonly used rates are 1% and $1\frac{1}{2}$% per month.

Last month Tom's credit card statement showed he owed $765.45. He made a payment of $200 and charged another $125.83 worth of merchandise. The credit card company charges 1% per month for finance charges. Estimate the amount of the finance charges that will be on this month's statement.

The actual amount is computed on the average daily balance, but an estimation can be made by finding how much Tom currently owes.

$$\$765.45 - \$200.00 + \$125.83 = \$691.28$$

Now find 1% of the amount owed.

$$\$691.28 \times 1\% = \$691.28 \times 0.01 = \$6.91$$

Practice

QUICK TIP

For large purchases, borrowing the money from a lending institution may cost less in interest than charging on a credit card and paying interest on unpaid balances.

Albert bought a set of golf clubs and paid $50 down and $26.25 per month for 12 months. The purchase price was $350.

1. How much did Albert pay in all for the golf clubs?

2. How much more than the purchase price did he pay under the installment plan?

3. What rate of interest was Albert charged?

Therese wants to borrow $3,500 to buy a used car. She does some research and finds she has two options.

4. She can borrow $3,500 on a level-payment plan from her bank and repay it in 6 monthly installments of $601.30. How much money will she pay in finance charges under this plan?

5. The car dealer offers a plan at 10% annual interest for 6 months. How much does this plan cost? (Hint: Since the payment period is only 6 months, use half of the 10% annual rate in your computation.)

6. Which payment plan costs less?

Answers: **1.** $365 **2.** $15 **3.** 4.3% **4.** $107.80 **5.** $175 **6.** The bank's plan

BALANCING YOUR CHECKBOOK

When the bank sends you the monthly statement for your checking account, it's a good idea to make sure your records correspond to the bank's records. If you know the balance in your checkbook is correct, you'll avoid unpleasant surprises. Reconciling your bank statement is easy if you've kept careful records. You must make sure you write *every* transaction in your checkbook.

Example

Lisa wrote each of her transactions in her checkbook register. When her bank statement arrived on March 24, she put a check mark against all the transactions that appeared on her statement, with this result:

NUMBER	DATE	DESCRIPTION OF TRANSACTION	PAYMENT /DEBIT (-)	√ T	FEE (IF ANY) (-)	DEPOSIT/CREDIT (+)	BALANCE $ 330 30
1344	3/10	Betty Smith	$ 20 00	$	$		310 30
1345	3/12	Van's Groceries	45 18	✓			265 12
1346	3/14	Midtown Garage	78 63	✓			186 49
	3/15	Automatic Teller	50 00	✓			136 49
	3/16	Deposit		✓		320.16	456 65
1347	3/20	Telephone Company	35 56				421 09
1348	3/22	Essex Dept. Store	196 24				224 85
	3/23	Deposit				50.00	274 85

Lisa noticed that checks numbered 1347 and 1348 did not appear, and neither did her $50 deposit on March 23. Furthermore, she had a service charge of $2.00. The monthly balance on the bank statement was $474.65.

To reconcile her account, Lisa first entered the $2.00 service charge in her checkbook and subtracted it from her balance, which left $272.85. Next she used a calculator to subtract from the bank's balance the checks she had written that didn't appear on the statement. Then she added the deposit that didn't appear on the statement.

Press: 474.65 − 35.56 − 196.24 + 50 =
The display reads 292.85. Lisa is $20.00 off.

Next she went further back through her check register and noticed that the $20 check she had written to

Betty Smith had not cleared. This was the source of the error. When she subtracted this $20 check from the bank's balance, Lisa's checkbook agreed with the bank statement.

Practice

1. Matt's checkbook register shows a balance of $306.56. His bank statement shows a balance of $387.12. His check register shows the following checks still outstanding: $100, $49.26, and $35.80. He has $100 per month automatically transferred from his checking account to his savings account, and he sees from his statement that he had forgotten to write it in his checkbook register and subtract it from his balance. In addition, Matt sees on his bank statement a $4.50 service charge.

Is Matt's bank statement correct? What is Matt's balance?

2. Jack got his bank statement in June and checked off the transactions that had cleared, with the following result:

NUMBER	DATE	DESCRIPTION OF TRANSACTION	PAYMENT /DEBIT (-)	√ T	FEE (IF ANY) (-)	DEPOSIT/CREDIT (+)	BALANCE $	
486	6/1	Logan Real Estate	$ 375 00	V	$	$	207	36
487	6/1	Electric Company	53 12	V			164	24
	6/3	Automatic Teller	75 00	V			89	24
	6/6	Deposit		V		365.00	454	24
488	6/7	Insurance Company	207 50				246	74

The check to the insurance company had not cleared. Jack has an interest-bearing checking account, and his statement says he was paid $3.11 interest. The balance on the bank statement is $447.35.

Jack adds $3.11 to his checkbook balance to get $249.85. He subtracts the one outstanding check for $207.50 from the balance on the bank statement, $447.35, and gets 239.85. Why does his balance not equal the bank's balance?

Answers: 1. Yes; $202.06 **2.** He made an arithmetic error in subtracting his check to the electric company. 207.36−53.12=154.24.

DOUBLING AND HALVING RECIPES

Recipes usually tell you the amount the listed ingredients will make—for example, four servings, two dozen cookies, and so on. Often you may want to make twice as much or half as many, so you must double or halve the ingredient amounts listed in the recipe. This means that you may need to multiply fractions by fractions.

Examples

The following recipe for baked potatoes stuffed with vegetables makes 8 servings.

4 baked potatoes
1 cup white sauce
$\frac{1}{4}$ teaspoon salt
$\frac{1}{2}$ cup grated cheese
$\frac{1}{2}$ cup each cooked peas and carrots
$\frac{1}{3}$ cup diced green pepper
2 tablespoons diced pimientos

Suppose you want to make 4 servings. How should you change the ingredient amounts?

Since you want to make $\frac{1}{2}$ as many servings, you should multiply each amount by $\frac{1}{2}$.

Potatoes: $\frac{1}{2} \times 4 = 2$ **potatoes**
White sauce: $\frac{1}{2} \times 1 = \frac{1}{2}$ **cup**
Salt $\frac{1}{2} \times \frac{1}{4} = \frac{1}{8}$ **teaspoon**
Cheese: $\frac{1}{2} \times \frac{1}{2} = \frac{1}{4}$ **cup**
Peas and carrots: $\frac{1}{2} \times \frac{1}{2} = \frac{1}{4}$ **cup each**
Green pepper: $\frac{1}{2} \times \frac{1}{3} = \frac{1}{6}$ **cup**
Pimientos: $\frac{1}{2} \times 2 = 1$ **tablespoon**

> ### QUICK TIP
> Remember that to multiply fractions, you multiply numerators together and denominators together. A whole number like 4 can be written as a fraction with a denominator of 1, for instance $\frac{4}{1}$.
> $\frac{1}{2} \times 4 = \frac{1}{2} \times \frac{4}{1} = \frac{1 \times 4}{2 \times 1} = \frac{4}{2} = 2$

The following recipe makes about 60 almond fortune cookies.

$\frac{3}{4}$ cup egg whites
$1\frac{2}{3}$ cups sugar
$\frac{1}{4}$ teaspoon salt
1 cup melted butter

1 cup flour
$\frac{3}{4}$ cup ground almonds
$\frac{1}{2}$ teaspoon vanilla

Suppose you want to double the recipe to serve at a party. How should you change the ingredient amounts?

Since you want to double the recipe, you should multiply each amount by 2.

Egg whites: $2\times\frac{3}{4}=\frac{6}{4}=1\frac{2}{4}=1\frac{1}{2}$ **cups**
Sugar: $2\times1\frac{2}{3}=2\times\frac{3}{3}+\frac{2}{3}=2\times\frac{5}{3}=\frac{10}{3}=3\frac{1}{3}$ **cups**
Salt: $2\times\frac{1}{4}=\frac{2}{4}=\frac{1}{2}$ **teaspoon**
Butter: $2\times1=2$ **cups**
Flour: $2\times1=2$ **cups**
Almonds: $2\times\frac{3}{4}=\frac{6}{4}=1\frac{1}{2}$ **cups**
Vanilla: $2\times\frac{1}{2}=\frac{2}{2}=1$ **teaspoon**

Practice

1. If a recipe calls for $2\frac{2}{3}$ cups of flour, how much flour should you use for $\frac{1}{2}$ recipe? (Hint: $2\frac{2}{3}=\frac{3}{3}+\frac{3}{3}+\frac{2}{3}=\frac{8}{3}$.)

2. If a recipe calls for $1\frac{1}{2}$ tablespoons of lemon juice, how much should you use for a double recipe?

3. If you double the following recipe for banana bread, how much of each ingredient will you need?

$1\frac{3}{4}$ cups flour
$2\frac{1}{4}$ teaspoons baking powder
$\frac{1}{2}$ teaspoon salt
$\frac{1}{3}$ cup shortening
$\frac{2}{3}$ cup sugar
$\frac{3}{4}$ teaspoon grated lemon rind
1 beaten egg
$1\frac{1}{4}$ cups mashed bananas
$\frac{1}{2}$ cup chopped nutmeats

Answers: 1. $1\frac{1}{3}$ cups **2.** 3 tablespoons **3.** $3\frac{1}{2}$ cups flour, $4\frac{1}{2}$ teaspoons baking powder, 1 teaspoon salt, $\frac{2}{3}$ cup shortening, $1\frac{1}{3}$ cups sugar, $1\frac{1}{2}$ teaspoons lemon rind, 2 eggs, $2\frac{1}{2}$ cups bananas, 1 cup nutmeats.

CALCULATING IN DIET AND EXERCISE

You can use math to help you plan a low-fat diet and to keep track of calories burned by various exercises. Using information given in charts, along with multiplying, dividing, and finding percentages, you can analyze your diet and decide on an exercise program.

Diet

Health experts now recommend that in a healthy diet, fat should account for no more than 20% to 30% of the total calories in the food we eat. You can figure the percentage of fat in your diet if you know the fat content and the number of calories in the food products you eat. As a rough guideline, you can figure that 1 gram of fat provides 9 calories of energy.

The product information chart on a breakfast cereal box indicates that one serving of the cereal provides 120 calories of food energy and contains 4 grams of fat.

A. To find the number of calories coming from fat, multiply 9 calories times 4 grams of fat.

9×4=36

B. Now find what percent 36 calories is of the total number of calories.

36÷120=0.3

Since 0.3=30%, you know that 30% of the calories in a serving of the cereal come from fat.

Practice

Use the information in the following chart for the problems.

Food	Fat grams	Calories
1 cup of whole milk	8	160
1 cup of skimmed milk	1	88
Fried chicken breast	14	175
Skinned, broiled chicken breast	5	100
3 chocolate chip cookies	6	150
1 apple	0	90

1. Compare the percentages of fat in one cup of whole milk and one cup of skimmed milk.

2. Compare the percentages of fat in fried chicken breast and broiled chicken breast.

3. Compare the percentages of fat in three chocolate chip cookies and one apple.

4. Suppose your daily calorie intake is 2,000 calories. How many grams of fat should be included in the food you eat if you want to keep the amount of fat-gram calories at 25% of your total calorie intake?

Exercise

Because exercising burns up calories, regular exercise, along with diet control, is an effective way to lose weight.

Fast walking burns up 280 more calories per hour than sitting still. Instead of watching television, if you walk one hour per day for five days a week, how many calories have you burned up?

To find the number of calories burned up, multiply the number of calories per hour by the number of hours.

280×5=1400

You have burned up 1,400 calories in a week.

Example

To lose one pound of weight, you must take in 3,500 fewer calories than you need to maintain weight or you must burn up 3,500 more calories than you use to maintain weight. How many days of fast walking for one hour per day would burn up 3,500 calories?

To find the number of days, divide 3,500 by the number of calories burned up per day.

3500÷280=12.5

It would take 12 or 13 days of fast walking to burn up 3,500 calories of energy.

Practice

Use the information in the chart below for the problems.

Activity	Calories burned per hour
Fast walking	280
Bicycling	330
Skating	420
Swimming	430
Jogging	570
Tennis	420

5. If you bicycle one hour per day, how many days will it take to burn up 3,500 calories of energy solely by bicycling?

6. How many more hours of fast walking than of swimming are required to burn up 3,500 calories?

7. Suppose you reduce your calorie intake by dieting to 500 calories per day below the number of calories you need to maintain your present weight. How many days will it take you to lose one pound?

8. If you swim one hour daily, along with the diet described in problem 9, how long will it take you to lose one pound?

Answers: 1. Whole milk, 45%; skimmed milk, 10.2% **2.** Fried chicken, 72%; broiled chicken, 45% **3.** Cookies, 36%; apple, 0% **4.** 55.6 grams **5.** 11 days **6.** about 4.4 more hours **7.** 7 days **8.** 4 days

COMPARING ATHLETES

In professional and college sports, players' performances are compared using a number of different statistics. In baseball, hitters and pitchers are rated using batting averages and earned-run averages. Basketball players are rated using points-per-game averages. Football carriers are rated on the average number of yards they gain each time they carry the football.

Earned-run average

One statistic used in baseball to compare pitchers is the earned-run average (ERA). Earned runs are runs scored by the opposing team as a result of a hit, a walk, a balk by the pitcher, or a sacrifice play. A pitcher's ERA gives an indication of the number of runs he might allow the opposing team to score during the course of a 9-inning game. To find a pitcher's ERA, first divide the total number of innings pitched by 9. This converts the total to a number representing complete 9-inning games pitched. Then divide the number of earned runs the pitcher allowed by the number of 9-inning games pitched to get the number of earned runs per game.

During the 1991 baseball season, Greg Maddux of the Chicago Cubs pitched 263 innings and allowed 98 earned runs, and Dwight Gooden of the New York Mets pitched 190 innings and allowed 76 earned runs. Which pitcher had the better ERA?

To find Maddux's ERA, first divide innings pitched by 9 to find games pitched.

$263 \div 9 = 29.2222$

Then divide earned runs allowed by games pitched.

$98 \div 29.2222 = 3.35$

Follow the same process to find Gooden's ERA.

$190 \div 9 = 21.1111$
$76 \div 21.1111 = 3.6$

Because a lower ERA is better than a higher one, Maddux had the better ERA in 1991.

Batting average

To find a baseball hitter's batting average, divide the number of hits the player gets by the number of times the player is at bat (walks and sacrifice outs are not counted). Batting averages are written with 3 digits to the right of the decimal point.

The National League's leading batter in 1991 had 187 hits in 586 times at bat. The American League's leading batter had 201 hits in 589 times at bat. Who had the higher batting average?

Divide hits by times at bat to find the averages.

National League leader: 187÷586=.319
American League leader: 201÷589=.341

The American League leader had the higher batting average.

Practice

1. Which pitcher has the best ERA?

Pitcher	Innings pitched	Earned runs allowed
John	254	87
Bob	219	100
Russ	275	92
Cal	283	90

2. Which hitter has the best batting average?

Hitter	Times at bat	Hits
Bill	548	181
Ted	546	179
Avie	611	195
Doug	620	210

Basketball scores

To find a basketball player's average number of points scored per game, divide the total number of points scored by the number of games played.

Which player had the higher scoring average in 1990–1991, Barkley or Robinson?

Player	Points scored	Games played
Barkley	1,849	67
Robinson	2,101	82

To find the scoring averages, divide the points scored by the games played.

Barkley: 1849÷67=27.6
Robinson: 2101÷82=25.6

Barkley had the higher scoring average.

Practice

3. Rank these players in order from highest to lowest scoring average.

Player	Points scored	Games played
Pat	2,154	81
Chris	2,107	82
Mike	2,580	82
Carl	2,382	82

Football yards gained

In football, players who receive the ball from the quarterback and run to gain yards are rated in terms of the average number of yards they gain each time they carry the football. You find this average by dividing the total number of yards gained by the total number of times the player carried the football.

Two of the all-time great ball carriers were Walter Payton and Jim Brown. Which player had the higher average number of yards gained per carry?

Player	Total yards	Total carries
Payton	16,726	3,838
Brown	12,312	2,359

Divide number of yards by number of carries to find the averages.

Payton: 16726÷3838=4.4
Brown: 12312÷2359=5.2

Brown had the higher average number of yards per carry.

Practice

4. Rank these ball carriers from highest to lowest in terms of the average number of yards gained per carry.

Player	Total yards	Total carries
Barry	1,304	255
Randall	942	118
Neal	1,078	260
Ernest	1,219	295

Answers: 1. Cal **2.** Doug **3.** Mike, Carl, Pat, Chris **4.** Randall, Barry, Neal, Ernest

READING CHARTS, GRAPHS, AND MAPS

Graphs and charts present information in a visual form to make it easier to understand. Furthermore, any assemblage of statistical data has been organized to convey a particular point of view; the graph or chart may be designed to emphasize certain trends or disguise others. When analyzing information presented in charts, graphs, or maps, carefully read all titles, scales, and legends before you begin. Be on the alert for information that is not included.

Charts

Voter Turnout in Presidential Elections			
Source: Committee for the Study of the American Electorate			
	1984		1980
	Registered Voters Voting	Voting-Age Population Voting	Voting-Age Population Voting
South Carolina	69.4%	40.6%	40.4%
South Dakota	71.8	63.8	67.3
Tennessee	66.4	49.3	48.7
Texas	68.3	47.0	44.9
Utah	74.9	60.5	64.4
Vermont	72.8	60.0	57.6
Virginia	84.1	51.1	47.6
Washington	76.3	58.5	57.4
West Virginia	71.8	51.3	52.8
Wisconsin	n/a	63.4	67.4
Wyoming	78.8	51.8	53.2

Which state in the chart had the highest percentage of voting-age people voting in the 1984 presidential election?

The question asks for information from the middle column, the percentages for all people of voting age. The state with the highest percentage is South Dakota, with 63.8%.

1. Which states showed an increase in percentage from 1980 to 1984?

2. Explain why the percentages in column 1 are so much greater than those in column 2 or 3.

Bar graphs

Bar graphs are sometimes used to emphasize relationships between groups. This one shows the number of days per year workers were absent due to a cold.

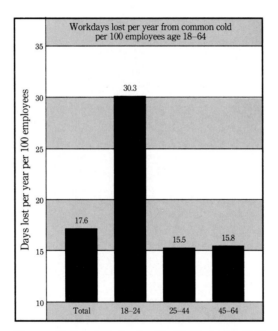

Source: National Center for Health Statistics, 1990

The leftmost column shows that for each 100 workers, 17.6 days were lost per year from the common cold. Which age group lost the most days per year? How is this indicated on the graph?

The tallest bar, reaching beyond the line for 30 days per year, belongs to the 18 through 24 age group.

Practice

3. Can you tell from the graph the total number of workdays lost per year from the common cold?

4. Judging from the relative heights of the bars, how many times as many days are lost because of a cold by workers aged 18 through 24 than by workers in the other two age groups?

5. What is the actual relationship between days lost by workers aged 18 through 24 and by the other two age groups?

6. How do you account for the difference between the answers for problems 4 and 5?

Maps

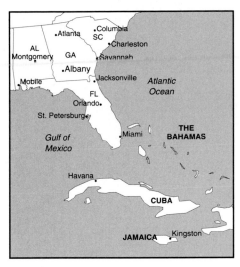

Use the map to estimate the distance from Jacksonville to Orlando. Measure the distance with an inch ruler. The distance on the map is about one-quarter of an inch.

If 1 inch=350 miles, 0.25 inch=? miles

To solve this, you need to know the relationship between 1 inch and 0.25 inch. You can tell by logic that 0.25 inch is $\frac{1}{4}$ of 1 inch; therefore the answer is $\frac{1}{4}$ of 350. The mathematical operation is to divide 0.25 by 1 and then multiply 350 by the result.

Use a calculator. Press: 0.25 ÷ 1 × 350 =
The display is 87.5.

The actual distance from Jacksonville to Orlando is about 90 miles.

Practice

Use the map above for the next two problems.

7. Which of the following cities is closest to Miami, Florida?

   a. Atlanta, Georgia
   b. Columbia, South Carolina
   c. Havana, Cuba
   d. Mobile, Alabama

8. Harry said he thought it was about 400 miles from Mobile, Alabama, to Orlando, Florida. Check his estimate on the map.

Answers: 1. South Carolina, Tennessee, Texas, Vermont, Virginia, Washington **2.** Not all people of voting age are registered to vote. Thus, the base number used to compute the percentages is smaller for column 1. **3.** No. The total number of workers is not given. **4.** Nearly three times. **5.** About double. **6.** The lowest number on the chart is not 0 but 10. If the lowest number were 0, the bars for 25–44 and 45–64 would be nearly $\frac{2}{3}$ longer, but the bar for 18–24 would be only about $\frac{1}{3}$ longer. **7.** c **8.** Harry's estimate is too high. The distance between the two cities on the map is less than 1 inch. The distance must therefore be less than 350 miles.

OPERATING YOUR CAR

Having a car generates two main categories of expenses. One category is fixed expenses that don't depend on how much you drive the car. The second category covers those expenses related to driving the car.

The chief fixed expenses are the cost of the car, license fees, and insurance. Figuring ownership expenses depends on whether you are financing the car or paying cash for it. If you pay cash for a car, you should spread the cost over the number of years you intend to drive it. If you buy a car on time, multiply your monthly payments by 12 to get the cost per year for the car.

Example

Althea has monthly car payments of $275. She pays $80 per year for state and local licenses, and she pays $325 semiannually for car insurance. What are Althea's yearly expenses for owning her car?

A. To find the annual amount for payments, multiply monthly payments by 12.

275×12=3300

B. Multiply the semiannual insurance payments by 2.

325×2=650

C. Add payments, insurance, and license fees to find the annual cost of car ownership.

3300+650+80=4030

Althea's yearly expenses for owning her car are $4,030.

Variable expenses

Car operating expenses include the cost of gasoline, repairs and maintenance, and parking and toll fees. These expenses can vary greatly, since they are affected by the amount of driving you do and by whether or not you must pay sizeable parking fees on a regular basis.

Example

Althea drives her car to work every day and uses it often on weekends. She uses an average of 10 gallons of gas per week. She gets the oil changed four times a year at an average cost of $17 per oil change. She takes her car for a checkup once a year at a cost of $165. She pays an annual fee of $100 to park her car in the company parking lot where she works. If Althea pays $1.29 per gallon for gas, what are her total annual operating costs for her car?

A. To find the cost of gasoline for a year, first multiply the gallons used per week by the price per gallon. Then multiply by 52 weeks per year.

1.29×10=12.90
12.90×52=670.80

B. Find the cost of four oil changes per year.

17×4=68

C. Now find the total for gas, oil changes, checkup, and parking.

670.80+68+165+100=1003.80

Althea spends $1,003.80 per year on operating expenses.

Practice

1. What are Althea's total expenses per year for owning and operating her car?

2. What are Althea's car expenses per month?

3. If public transportation to and from work costs you $3.00 per day, how much would you pay in one year to travel to and from work? (Assume 50 weeks of work per year.)

Answers: **1.** $5,033.80 **2.** $419.48 **3.** $750

RENTING VERSUS OWNING

Should you rent or should you buy? In most cases, buying a house is a good financial investment. The value of a house tends to increase over the years, and homeowners have federal and state tax advantages. But buying a house usually requires a substantial amount of money for a down payment, and homeowners have a number of expenses—interest on the mortgage, real estate taxes, homeowner's insurance, structural repairs, replacement of major appliances, and heating bills, for example—that apartment dwellers don't usually have. How does the cost of owning compare with the cost of renting?

Example

The Harrises rent a three-bedroom apartment; the Jacksons own a three-bedroom house. The chart below shows their housing expenses for last year. (The initial down payment is not considered in this example.)

Harrises
Rent: $1150 per month
Personal property insurance: $350 per year
Garage rental: $75 per month
Phone and electricity: $95 per month

Jacksons
Mortgage principal and interest: $850 per month
Homeowner's insurance: $400 semiannually
Heat: $275 per month
Phone and electricity: $105 per month
Real estate taxes: $2,500 per year
Roof repair: $1,500
New hot-water heater: $650

To compare the two families' housing expenses, figure the total amount each spent last year.

A. Figure the Harrises' expenses.

Rent: 12×1150	**$13,800**
Insurance	350
Garage rental: 12×75	900
Phone and electricity: 12×95	1,140
Total	**$16,190**

B. Figure the Jacksons' expenses.

Mortgage: 12×850	$10,200
Insurance: 2×400	800
Heating: 12×275	3,300
Phone and electricity: 12×105	1,260
Real estate taxes	2,500
Roof repair	1,500
Hot water heater	650
Total	$20,210

C. Subtract $20,210−$16,190 = $4,020

The Jacksons' expenses for housing were about $4,000 more than the Harrises'.

Practice

1. The Jacksons' mortgage payments last year included $4,800 in interest. They are allowed to deduct their interest payments and their real estate taxes from their adjusted gross income for federal tax purposes. What is the total deduction for these two items?

2. If the Jacksons' federal income tax is 31% of their adjusted gross income minus deductions, how much will the two deductions listed in problem 1 save them in taxes?

3. If you subtract the Jacksons' tax savings from their housing expenses, how did their expenses compare with the Harrises' expenses?

4. The Harrises have an annual spendable income of $50,000. What percentage of their income is going for housing?

5. The Jacksons have an annual spendable income of $52,000. What percentage of their income is going for housing?

Answers: 1. $7,300 2. 31% of $7,300, or $2,263 3. About $1,700 more 4. 32% 5. 39%

CONSERVING ENERGY

Two ways to conserve energy and, in the process, save money are by driving a car that gives high gas mileage and by heating or cooling your home with a unit that has a high energy efficiency rating.

Gas mileage

When Alice last bought gas, her odometer reading was 32042. She just bought 12.8 gallons of gas to fill her tank, and her new odometer reading is 32362. How many miles did she get per gallon?

A. Subtract 32042 from 32362 to get the number of miles driven.

32362−32042=320

B. Next divide the number of miles driven by the number of gallons it took to fill her tank.

320÷12.8=25

Alice got 25 miles per gallon.

> ### QUICK TIP
> **If you have a trip odometer, set it when you fill up. The next time you buy gas, you'll have a reading of the miles you've driven.**

Energy efficiency

Newer models of appliances tend to have higher energy efficiency ratings than older models. They use less energy and cost less to operate. Many new appliances have energy guides attached.

The cooling capacity of room air conditioners is measured in Btuh (British thermal units per hour). They are given an energy efficiency rating (EER), which is calculated by dividing the Btuh by the number of watts of power used. The higher the rating the fewer watts are used relative to cooling power, so higher ratings are more efficient.

A. If a 6,000-Btuh air conditioner uses 630 watts of power, what is its EER?

$$\frac{\text{Btuh}}{\text{watts}}=\text{EER}$$
6000÷630=9.52

The air conditioner has an EER of about 9.5.

B. Another air conditioner has 5,400 Btuh of cooling power and has an EER of 9.0. How many watts of power does it use?

$$\frac{\text{Btuh}}{\text{EER}} = 9.0$$
$$5400 \div 9 = 600$$

The air conditioner uses 600 watts.

Example

The following chart is from a 22,500-Btuh air conditioner with an EER of 8.5.

Yearly hours of use	250	750	1000	2000
	Estimated yearly cost			
Cost of electricity 6¢	40	119	159	318
per kilowatt 8¢	53	159	212	424
hour 10¢	66	199	265	529
11¢	79	238	318	635

Suppose this air conditioner runs for 750 hours in one year and electricity costs 8¢ per kilowatt hour. To find the cost to operate this air conditioner for the year, look for 750 hours across the top row. Then read down that column to the amount in the same row as a cost of 8¢ per kilowatt hour. The amount is $159.

Practice

1. Ali gets 37 miles per gallon. Sue gets 23 miles per gallon. Each drives about 12,000 miles per year. If gas costs $1.35 per gallon, how much less does Ali spend in one year for gasoline?

2. The air conditioner in the example replaced an older unit that used 3.2 kilowatts of electricity per hour. What was the total cost of running this unit for 1,000 hours at a cost of 10¢ per kilowatt hour?

3. For 1,000 hours of usage at 10¢ per kilowatt hour, how much was saved in one year?

Answers: 1. $266.51 2. $320 3. $55

59

PLANNING HOME IMPROVEMENTS

Home improvement projects often require calculations with measurements. You might need to change measurement units from inches to feet, for example, or from square feet to square yards. Furthermore, a wise do-it-yourselfer needs skills in interpreting information from the charts found in catalogues or in consumer information articles.

Perimeter

The simplest type of math used in home projects involves linear, or straight-line, measurements. When you find the perimeter of (distance around) a space, you are using linear measurements.

Art has a small lot on one side of his house. He wants to fence in the lot to keep his dog from rushing out into the street. The lot extends 20 feet away from the house and is 10 feet deep. How many linear feet of fence does Art need?

Art needs fencing for only three sides of the 10-by-20-foot rectangle. The fourth side of the rectangle is formed by the house.

Add the three dimensions:

20 feet+10 feet+20 feet=50 feet

Art needs 50 feet of fence.

Practice

1. How would the answer to the problem in the example change if the house were not there?

2. Bonnie's bedroom measures 10 feet by 15 feet. She decides to put up a flowered wallpaper border. If the border is to go all the way around the room, how many feet does she need?

Area

Jack sees an ad for roofing shingles at $5.79 per bundle. The ad says that 3 bundles cover 100 square feet. The two sides of Jack's roof are each about 80

QUICK TIP

The answers to your calculations can be only as good as your measurements. A common rule of thumb is "measure twice, cut once."

feet by 60 feet. How many bundles of shingles will he need?

Jack's roof has two parts, each 80 feet by 60 feet. The total area is thus twice the product of 80 and 60.

$2 \times 80 \times 60 = 9600$ square feet

Dividing the total area by 100 gives 96 sections of 100 square feet each. Jack will need 3×96, or 288 bundles.

Practice

QUICK TIP

Using squared paper to make a rough scale drawing of a room or apartment can make planning easier. A convenient proportion is 4 or 5 squares to the inch. Use the scale of 1 square on the paper equals 1 square foot.

3. In the example above, how much will Jack need to spend for the shingles?

4. Sharon plans to put insulation in her 45-by-60-foot attic. She chooses insulation that is $3\frac{1}{2}$ inches thick. Each 50-square-foot roll sells for $8.99.
   a. What is the area of Sharon's attic in square feet?
   b. How many rolls of insulation will she need?
   c. How much will the insulation cost?

5. Sam reads in a consumer article that the coverage for interior semigloss paint is 650 square feet per gallon. Sam plans to paint a room 10 feet long and 15 feet wide, with an 8-foot ceiling.
   a. Find the total number of square feet for all 4 walls of the room Sam plans to paint.
   b. Sam thinks the room will need 3 coats of paint. How many gallons of paint should he buy?

Answers: **1.** The perimeter would be 10 feet greater—60 feet in all. **2.** 50 feet **3.** $1,667.52 **4.** a. 2,700 square feet b. 54 rolls c. $485.46 **5. a.** 400 square feet **b.** 2 gallons

CONVERTING UNITS OF MEASURE

For some home improvement projects, you may have to convert from one unit of measurement to another.

Example

Carol uses a yardstick and finds that the dimensions of her bathroom are 2 yards wide and 4 yards long. She would like to install a ceiling ventilator recommended for rooms up to 65 square feet. Is the ventilator big enough for her bathroom?

One way to proceed is to calculate the number of square yards and convert to square feet by multiplying by 9. You can also convert 65 square feet into square yards by dividing by 9.

65 square feet÷9=7.2 square yards

The area of the bathroom is 2×4, or 8 square yards. The ventilator may be too small for the room.

Practice

Many types of floor tile are smaller than 1 square foot. Computing the number of tiles needed for a floor requires changing square feet to square inches.

1. Ruth is going to use the mosaic tile for her kitchen floor. The floor is 13 feet by 10 feet. How many packs of tiles does Ruth need to buy?

2. How many of the 8-by-8-inch floor tiles are needed for a 10-by-12-foot kitchen floor?

3. The 8-by-8-inch tiles come in 11.11-square-foot packs. Find the number of tiles in each pack.

U.S. System

Length
12 inches=1 foot
3 feet=1 yard
36 inches=1 yard
5280 feet=1 mile
1760 yards=1 mile

Area

144 square inches=1 square foot
9 square feet=1 square yard
4840 square yards=1 acre

Volume

1728 cubic inches=1 cubic foot
27 cubic feet=1 cubic yard

Weight

16 ounces=1 pound
2000 pounds=1 ton

Capacity

8 fluid ounces=1 cup
2 cups=1 pint
2 pints=1 quart
4 quarts=1 gallon
32 quarts=1 bushel

Metric System

Length

10 millimeters=1 centimeter
100 centimeters=1 meter
1000 meters=1 kilometer

Mass (Weight)

1000 milligrams=1 gram
1000 grams=1 kilogram

Capacity

1000 milliliters=1 liter
1000 liters=1 kiloliter

Answers: **1.** 13 packs **2.** 270 tiles **3.** 25 tiles

CONVERTING METRIC UNITS

The United States is the only major country to still use the foot, quart, and pound. Almost all other countries use metric units of measure. Because of foreign trade, many items bought in the United States are labeled with metric units. When shopping and traveling, it's necessary to convert metric units to U.S. units.

Example

> **QUICK TIP**
>
> You can convert kilometers into miles easily if you remember that 8 kilometers equals 5 miles.

Canada uses the metric system. Instead of a speed-limit sign reading 65 MPH, what would you expect to see on a Canadian speed-limit sign?

Since 1 mile=1.61 kilometers, multiply 1.61 by 65 to find an equivalent speed in kilometers.

65×1.61=104.65

You would expect Canadian speed-limit signs to read 100 KPH or 105 KPH.

Metric Equivalents

> **QUICK TIP**
>
> 1 foot equals about 30 centimeters.

1 inch=2.54 centimeters
1 foot=0.30 meters
1 yard=0.91 meters
1 mile=1.61 kilometers

1 ounce=28.35 grams
1 pound=0.45 kilogram

1 fluid ounce=29.57 milliliters
1 cup=0.24 liter
1 pint=0.47 liter
1 quart=0.95 liter
1 gallon=3.79 liters

0° Fahrenheit=−17.8° Celsius
32° Fahrenheit=0° Celsius
212° Fahrenheit=100° Celsius
1° Celsius=$\frac{5}{9}$° Fahrenheit